آكاٽي نمبرجي

THE NUMBER STORY

SMALL BOOK ONE

ENGLISH - SINDHI

Numbers Teach Children
Their Number Names

written and illustrated by

MISS ANNA

Early Reader Edition of *The Number Story 1*
Bronze Medal Winner, 2016 Wishing Shelf Book Award

Library of Congress Control Number: 2018902040

Names: Miss Anna, author.
Title: Number story : numbers teach children their number names / Miss Anna.
Description: Portland, OR: Lumpy Publishing, 2018.
Identifiers: ISBN 978-1-949320-10-7 | LCCN 2018902040
Summary: The pictures and rhymes present stories which introduce numbers 0-10.
Subjects: LCSH Numeration—English—Sindhi--Pictorial works--Juvenile literature. | BISAC JUVENILE NONFICTION /
Languages: English—Sindhi
Classification: LCC QA141.3 .M57 2018 | DDC 513—dc23

Publisher: Lumpy Publishing
Website: www.missannabooks.com
Email: missanna@missannabooks.com

Paperback: ISBN 978-1-949320-10-7
Printed in the U.S.A.   1 3 5 7 9 10 8 6 4 2

اسان جي نمبرن جا نالا سکڻ چاهيو ٿا؟

It is very easy and a lot of fun!

هي ڏاڍو آسان آهي ۽ ان ۾ ڏاڍو مزو آهي!

Say-along our little jingle

اسان جي ننڍڙي آکاڻي سان ڳايو!

starting from Number One!

اسان نمبر هک سان شروع کنداسين!

1

ONE looks like my one finger.

هي منهنجي هڪ آڱر وانگر سڌو آهي.

ONE!

هكا!

2

TWO trails a tail.

ب

ُچ جھڙو نشان.

A TAIL! ‏هڪ پُڇ!‏

3

THREE   has bumps.

ٽي

هي هڪ ٽڪري وانگر آهي.

BUMPY!

سايون تکريون ڎسو!

4

FOUR carries a sail.

چار

بادبان لڳل آهي.

A SAIL!
بادبان!
بادبان سان هڪ ييڙي!

5

FIVE   is a racing track.

پنـج

هـي هڪ ريسنگ ٽريڪ آهي.

VROOM
وروم
1

6

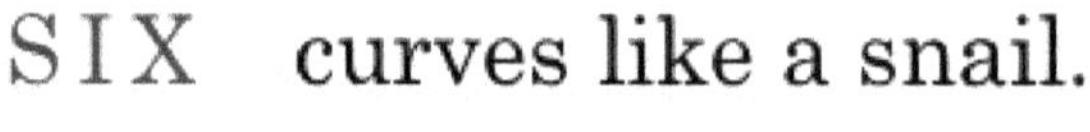

SIX   curves like a snail.

چھ

گھونگھي وانگر موڙ.

A SNAIL!   ‫ده گهونگهو!‬

7

SEVEN has a sharp angle.

ست

تكو زاويو آهي.

BE CAREFUL! IT'S SHARP!

سنيال كر!اڎايو تكو اَهي!

8
EIGHT   is rollercoaster rails.
ات
رولر کوسٹر آهي.

يپي!
YIPPEE!

9

NINE is a bubble on a stick.

نَئون

كائتي تي بُلبلو.

A BUBBLE!

هك بُلبلو!

10

TEN is an eye of a whale.

هـي ويل جي اک آهي.

هيلو!

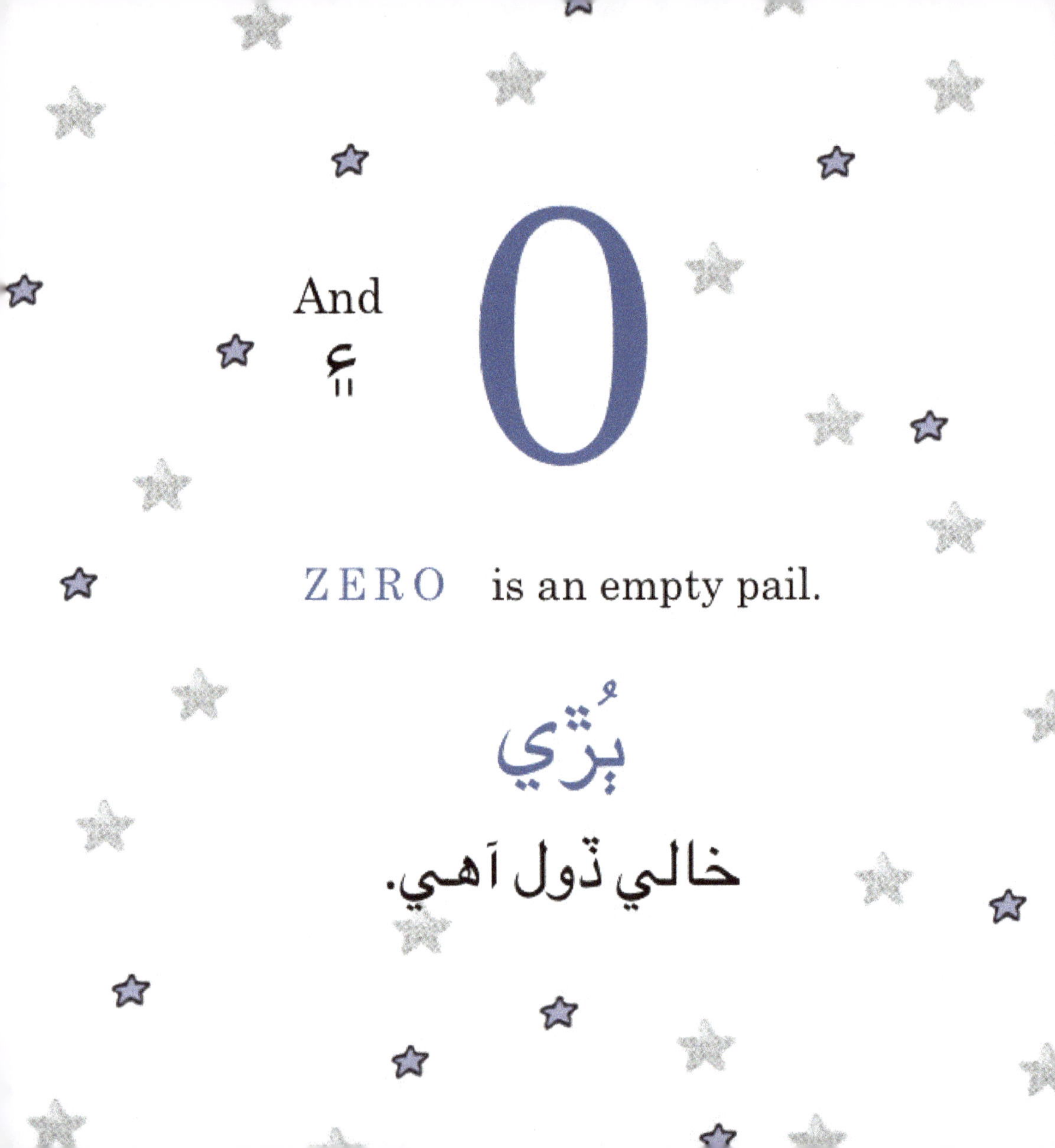
And

O

ع
ZERO   is an empty pail.

يُژي

خالي ڏول آهي.

IT'S
EMPTY!

هي خالي آهي!

Thank you for playing with us today.

We had a lot of fun too!

اڄ اسان سان کيڏڻ تي توهان جي مهرباني.
اسان کي به ڏاڍو مزو آيو!

We are your Number friends,
Zero to Ten,
Who will be here for you~

اسان توهان جا دوست آهيون
پُرّي کان ڏه تائين.
اسان هتي سدائين توهان لاءِ هونداسين.

Bye-bye now!
See you again soon!
في الحال بائي بائي!
توهان سان ڀيهر جلد ملاقات ٿيندي!

The Numbers are *SINGING* too!

To sing-a-long, look for Miss Anna Number Story
at your favorite music store like iTUNES.

MP3

Numbers 0-10
IDENTIFYING & COUNTING

Number Story 1 & 2

isbn: 978-0-996216-48-7

Numbers 11-20 & Ordinals
first, second, third...

Number Story 3 & 4

isbn: 978-1-945977-01-5

Numbers 0-100 & Place Values
ones, tens, hundreds...

Number Story 5 & 6

isbn: 978-1-945977-06-0

About Clocks & Telling Time
hours, minutes, seconds

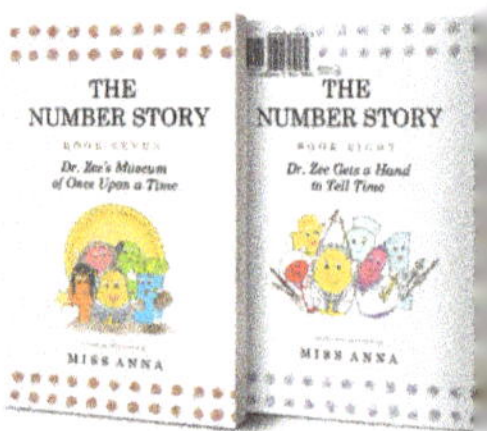

Number Story 7 & 8

isbn: 978-1-949320-40-4

For more Miss Anna books to love,
visit us at

www.missannabooks.com

Numbers are working hard all over the world!
Come Travel the World with Us!